Leo H. Grindon

The Phenomena of Plant Life

Leo H. Grindon

The Phenomena of Plant Life

ISBN/EAN: 9783337095383

Printed in Europe, USA, Canada, Australia, Japan

Cover: Foto ©berggeist007 / pixelio.de

More available books at **www.hansebooks.com**

THE PHENOMENA

OF

PLANT LIFE,

By LEO H. GRINDON,

LECTURER ON BOTANY AT THE ROYAL SCHOOL OF MEDICINE, MANCHESTER; AUTHOR
OF MANCHESTER WALKS AND WILD FLOWERS, ETC.

BOSTON:

NICHOLS & NOYES.

1866.

THE delightful series of papers which compose this little volume are taken from recent numbers of an English periodical, " *The Intellectual Repository*."

We are led to collect and publish them in this form, from the great pleasure their perusal has afforded ourselves.

They will be found to describe, in beautiful language, the still more beautiful, and living " *Phenomena of Plant Life;*" and, if we mistake not, will afford a rich mental and moral repast to every careful reader.

Referring mainly to the *character* of the work, we doubt if a more elegant gift volume will be offered the present season.

PUBLISHERS.

Phenomena of Plant-Life.

CHAPTER I.

WINTER.

THE new year opens very appropriately in the depth of winter, since the commencement of all things, both in the natural and in the moral world, takes place in secrecy and seeming darkness. Yet new year's day is a matter only of the artificial division of time. The phenomena of living nature which mark the actual progress of the year, are independent altogether of the almanac. Long before we exchange our kindly greetings, and those happy salutations and generous wishes of the season, that signalize the advent of the new year to our firesides, — long before this, it has been new year's day to a thousand buds and flowers, both in field and garden. Delicate looms have been set in motion to weave that sweet apparel with which, in due time, even Solomon, in all his glory, might not compare. Deep in the hidden chambers of many a root and little bulb, have commenced in quiet energy those wonderful preparations which, when summer

1*

bids welcome, charm our eyes with lovely colors, and
our nostrils with aromas.   In a word, though to civil-
ized man it is the *first* day of the year, to vegetable
life it is a period of advanced infancy.   Rightly to
esteem the flow of the seasons, we must view them as
an unbroken sequence of new developments.   Though
one class of appearances may come to a close, another
rises out of it almost before we miss the departing
one ; as on a fair midsummer's night, before we have
lost the last trails of the reluctant sunset, the calm,
still, sweet aurora of the new sunrise, peeping over
the mountain-tops, enters our hearts like the smile of
a child.

In a word, again, we never see *beginnings*.   We
think we trace rivers to their sources, but the first
trickles among the moss on the mountain-side, are
collections of water-drops that have their own anterior
history.   The coy sources of the Nile, that have at
length rewarded enterprise, far back as they lie in
those sultry African plains, do but represent a stage
in the life of the immortal stream.   The forest that
has been venerable for ages, began in acorns and tiny
seeds, whence derived, even the philosopher can only
guess.   The shells that inlay the wrinkles on the
sands, — these come tossed up, it may be, from some
birthplace that human eye has never beheld ; — it is
always something in a measure *accomplished* that we
obtain ; early as we commence our search, we always

enter late, — the year has begun before we thought, or could be quick enough to watch. So it is with the operations of Divine Love. Everywhere we are steeped in blessings that lie back beyond all memory of beginning, or perception of cause. We may learn to appreciate more fully, — and understanding better, to be more grateful, but for the first flow of them, we must ask of the "morning stars" that "sang together," and of the "sons of God" that "shouted for joy." The simplest throb of pleasure that swells the soul in connection with the good or true, if we will but look at ourselves in the light of the recipients, that we are, is no incident purely of the hour, but a result of something our diary does not record; far, far away in the heavenly era of earliest boyhood, was sown the seed that brings forth that pleasant fruit.

Take first, as an illustration of this wonderful winter-life of plants, the little bulb of the common garden crocus. At this season, if we have not one at hand to dig out of the ground, it is easy to procure an example from any seed-shop. The bulb itself is round, flattened at top and bottom, and covered with elegantly-netted brown coats. Upon the summit are elevated several white spires, plump, hard, and pointed, and in these, if we dissect carefully, will be found all the golden glory that would have been unfolded in March and April. The petals are there, minute it is true, but in that respect not inferior, in their degree, to

kings and princes as they lie in their cradles ; the stamens are fully formed, and stand as the principal part of
the blossom, and round about are tiny spear-like leaves.
Every cluster is wrapped separately in transparent
clothing, and over the whole are strong and opaque
vestments that protect the precious rudiments alike
from cold and wet. By degrees the spires grow taller ; presently they burst at the tips, and eventually
the foliage and yellow vases peep above the ground.
The bees are glad when they arrive, and visit them al-
ternately with the palm-bloom in the hedges, returning from their happy labor all besprinkled with the
yellow pollen. If a few crocus bulbs be placed in a
tea saucer, with a little cotton-wool as a foundation,
and the saucer be kept constantly supplied with water,
so that the wool shall be permanently saturated with
wet, the spires will open just the same as if in the
earth, and make even the gloomiest of back sitting-
rooms cheerful at the dreariest season of the year,
opening their gay corollas one after another. To
watch them grow day by day, is alone a cheerful
sight. The more we can keep ourselves face to face
with the simple and pretty little things of nature,
bringing them into our parlors, nursing them upon
our mantelpieces, making them companions of our sol-
itude, the more truly do we learn to love what is
grand and noble in the outer world. Improving ideas
are not got only — nor perhaps so much — from the con-

templation of waterfalls, mighty mountains, and extended prospects, as from the day-by-day quiet observation of the wonderful ways of God in the calling
forth of a little flower from its nest, and painting it
with miraculous hues that seem impossible to proceed
from dull, cold soil. The glory of God is to bring order out of confusion, peace out of discord, life out of
death; and nowhere in nature do we see it more beautifully expressed than in the birth of the silver-mantled
flower, — a birth that comes not through any aid or
encouragement from man, but apparently of its own
free action. The yellow crocus is a·native of the
South of Europe, though introduced so long ago to
our own island, as to be one of the oldest inhabitants
of the garden. Few think, perhaps, when surveying
its varied loveliness, how many countries and how
many years of diligence have contributed to the garden; yet there is scarcely a country beneath the sky
but has been laid under contribution, and there is now
in England a summary of the floral treasures of the
whole world. The purple crocus, on the other hand,
is an ancient Briton, counting itself as part of the
grass of the field, — not, indeed, as a common object
of the country, like the primrose or the daffodil, but
as one of the select few that are confined to certain
spots. Near Nottingham, in March and April, the
meadows are flooded with its refreshing bloom, and
the flowers, brought home in handfuls by the city

children, lie scattered about the streets as though Nature had lost her way.

What the crocus during winter is in the earth, the flower-buds of many trees are upon the boughs. These very trees, which to the eye are least provided with flowery charms, and which never aspire, even in the height of their life, to be more than what neutral tints are to the artist, — these very trees are in winter so rich in wonder, as to take rank with the most alluring forms of nature. The common hazel-nut in mid-winter is hung with innumerable gray-green clusters ; the alder and willow-buds swell with leafy effort; the latter often burst before Christmas, and disclose their silky contents. Everywhere there is the note of preparation, and though the cold days and colder nights may chill their sap, the movement is still upwards ; — spring is the desire, spring is the promised land ; and though the fireside may prove more inviting than the woodland, and incuriousness may leave them all unobserved, no matter ; every tree moves its steady way, seeking outlets at a thousand points ; and by-and-by, when a tempting afternoon carries our footsteps across the meadows, we look round in congratulation that spring is beginning, whereas in truth, it is we who are just beginning to observe. Thus is winter in connection with plant-life, if we will only go forth and learn, a time of grand assurance to us that nothing ever absolutely ceases. The particular organs constructed for the performance of given functions may collapse and go to decay, but the life which acts through them never ceases for one instant. Sleep in the animal, leaflessness in the plant and tree, indicate only that nature is gathering up her strength for new movements ; that which seems cessation is the transit from a weaker to a more powerful state. Winter, in fact, is the necessity of all beginning, as summer is the necessity of all ripeness and perfection.

I have often been struck in winter by the peculiar beauty, then revealed, in the architecture of the oak as compared with the poplar, of the elm as compared with the larch, and so on, all through the long list of the vegetable patriarchs. Winter is needed in order that we may have their various figures truly disclosed, since in summer all is concealed by masses of foliage ; and it is not the least among the many solaces of drear December, that the manly dignity of one kind of tree shall be brought into contrast with the feminine gracefulness of another, and that we shall be reminded by these disclosures, that truth abides not in apparel, but in those inner lineaments of things which in the heyday of excitement and pleasure, we are apt to forget. In summer, we overlook ; the glory of the world encircles us, and we are content ; in the summer of life, similarly circled by its charms, we are as apt to forget that all is passing away ; we eat and drink and are merry. Thanks, then, be to God, that secular

pleasures fade as a leaf, that the outsides of things are
stripped away, and that the mournfulness of separa-
tion and bereavement come round inevitably, for these
are the processes that place us in the presence of what
is permanent. As winter in the natural world is to
the accomplished mind no time of gloom, but a period
rather for realizing new delights, though possibly
tinged with seriousness, so winter in the life of the
soul, need bring no despondency or sadness, since it
is then that we gather our best glimpses of immortal
truth.

So in winter have I often been gladdened by the
sight of the glorious ivy boughs, that mantling the
aged trunk, wreathe it with perennial and shining
verdure. When most other things are withheld, the
ivy, the holly-tree with its scarlet bracelets, the mis-
tletoe loaded with pearls, maintain for us the sweet
influence of nature, — images of indestructibility ; and
triumphing over darkness and cold, are well used to
decorate our houses and churches at Christmas. There
is more than appears at first sight in the use of these
plants for Christmas ornaments. Antiquaries refer us
to fancies of the ancients, that the sylvan deities
(themselves purely fabulous beings) being frozen out,
or at least benumbed, in their native woods, were glad
to take shelter, like robins, in the vicinity and beneath
the roofs of human habitations, and that these cheerful sprays of evergreens were to give them a kind of

welcome, so that they might not feel altogether lost
and exiled. The truth lies probably in the ancient
symbolic use of trees and shrubs in connection with
religious faith, on which ground they were used in
pious ceremonies, and placed beside the altars, as visible representatives of those peculiar blessings which
the deity to whom they were consecrated, was be-
lieved willing to bestow where reverently asked. Our
Lord's coming to the earth in the depth of winter, was
representative of the time when man most needs Him ;
these rich, red berries and lustrous leaves of the holly,
triumphing over the asperities of frost and snow, pic-
ture beautifully His dominion over the powers of darkness, and are life, as it were, made visible.

Nor are evergreens all that greet the eye in mid-
winter. There are flowers, too, few it may be, but
choice and pure, sufficient to assure us that Flora never
forgets her prime duty ; and though she may repose
awhile, provides sweetness for every day. Nothing
is fairer than the Christmas-rose, though not a rose in
nature. The large white petals form elegant concavi-
ties, with a tuft of yellow stamens in the centre, after
the fashion of a little wheat-sheaf, and round about the
latter is set a ring of green honeycups, in which, even
now, scanty nectar may be discovered. The appearance of these Christmas flowers, following as they do
many of late autumn, illustrates beautifully that although our seasons in the North are so marked, yet,

2

as in the tropics, where the procession is continuous, we in England have in reality all four periods side by side. Nature has so ordered her economy, that spring, summer, autumn, winter, are all represented every week in the year; not necessarily in atmospheric phenomena, but in the life-history of trees and plants. The little cresses that dot the wheat-fields with white flowers in March and April, have run through their summer and closed their autumn, while to the lilies it is only spring;—when the lilies stand white, and tall, and fragrant in their queenly pride, the Michaelmas-daisy and the farewell-summer are only bestirring themselves; and these, in their departure, are soon forgotten in the chrysanthemums. And so it goes on, life ever treading upon the steps of decay, all forces and phenomena summed up in every circle in which we may find ourselves. There is no retiring from the presence of life, any more than, by taking the "wings of the morning," that we may dwell in the "uttermost parts of the earth," we escape from the Presence that ruleth all.

CHAPTER II.

THE essential sign of spring in northern latitudes is the swelling of the buds upon the trees, and those of the sturdy bushes which the husbandman uses for hedges. The appearance of flowers, except to the experienced eye, cannot always be depended upon. Many that would be thought heralds of the new season are in reality relics of the year that has departed, — epitaphs on the summer of six months before, — memorials rather than prophecies. Such is the case with the wall-flower, which is often seen plentifully in bloom in January, unless the winter be very severe, — the succession of flowers from side-shoots having proceeded uninterruptedly perhaps since the previous May. This long-protracted flow of bloom is usually attributable to the flowers being gathered for love-tokens or personal pleasure, and thus hindered from fulfilling the grand purpose for which all flowers are in every case developed, namely, the origination of seed from which new plants shall be reared, to take the place of the parents, when the latter lie withered and dead. As long as a plant is hindered from pro-

ceeding with the due preparation of its intended
seeds, so long will it persist in its efforts, and renew
them, striving, till all its vitality is exhausted, to leave
if it be only a single voucher of its honest toil. A
thousand times have I noticed this wonderful and quiet
energy in operation. In the fields some hungry quadruped bites off the young green flower-head as a relish
to the insipid grass; — no matter, from every joint
below, a new shoot is soon put forth; and in a few
weeks, where there would have been, perhaps, no
more than a single blossom, there are now a dozen
flowers. So in the garden some lily hand crops a
flower white as itself, and if the structure of the plant
permit, by and by the whiteness gleams from one little
side branchlet after another, and in a way that would
probably never have happened save for the destruction
of the first-born. Applying our knowledge of this
principle to the interpretation of the Christmas wall-
flowers, it is easy to understand how it happens that
their bloom lingers so long. Many a posy, when the
days were at their best, was probably made odorous
with the early blossoms of this cheerful plant; these
that come in the dull cold days of winter are the proof
of the hindered efforts, and a witness to unflinching
perseverance in the fair endeavor, — a perseverance
that may read us all a gentle lesson, — strive to the
last; if we fail, we have at least deserved to win.

Very different are such flowers as the yellow pile-

wort and the golden coltsfoot. These are genuine
spring blossoms, never appearing until the new year
has made a fair start, nor renewing their flowers in
summer and autumn. Being "weeds," and never
growing in pastures, they are seldom cropped or
gathered, so that the original preparation of bloom is
generally followed by successful result in seed. Very
pretty is it, when the last of the snow has dissolved
from the ground, to see the bright rays of the pilewort
among the half-withered relics of the past autumn
upon the hedgebank, and their young leaves spreading
a carpet over the heretofore brown earth in woods and
groves; no less pretty is the spectacle of the coltsfoot, when it opens its yellow disk, formed of a hundred rays as fine as needles, and this without a single
leaf to stand in contrast. Both flowers need the
sunshine in order that they may expand. On dull and
cloudy days they remain fast shut up, but with the
first kind beam from the sky they spread their little
petals, and glow as long as the atmosphere is genial.
The pilewort is not unlike a buttercup, but the leaves
are rounded and polished, and it rarely grows taller
than the breadth of one's hand. "Weed" it may be
in popular estimation, but the wood-pigeons do not so
lightly esteem it. The fleshy roots, shaped like little
round beans, lie very near the surface of the soil: the
rain washes the earth from them, and lays them bare,

2*

and these birds come and make their meals on the
supply thus provided.

Every living creature has its cornfields; true, it is
only man who is called upon to sow and reap, to grind
and to bake into bread, — and this, in order that by
virtuous and regular labor he may have his intellect
and affections stimulated; but cornfields, in their kind,
are spread for everything that eats, — composed, it
may be, of the simplest and weakest plants in nature;
still, in their importance to tens of thousands of
speechless creatures, no less momentous than the
broad acres of wheat and barley, oats, rice, rye,
millet, and maize, that supply the human population
of the earth with their daily sustenance. There is
probably no plant in nature that does not directly
support the life of some little animal: it was for this
purpose that plants were in great measure called into
being, and when we are tempted to despise the insig-
nificant ones, and to call them "weeds," we should
remember that nothing has been made in vain, and
that everything has been designed for some generous
purpose. But why should they be called "weeds"?
Weeds are flowers out of the place for which Providence designed them. If a lily spring up by some
casualty in a potato-bed, it is in that place a weed,
quite as much so as a dandelion is among the tulips;
but neither of them is a weed in its native woods or
fields, since these are the habitations assigned to

them, and in which, to eyes that look on the sweet simplicity of creation with joy and pleasure, they are always beautiful. Very much of what we are apt to consider the uncomeliness of things comes in reality of our not seeing them in their natural and proper conditions, but under some artificial and constrained circumstances that interfere grievously with the native characters. Look, for instance, at the unfortunate monkey, dragged from its native haunts, and carried about the streets on an organ-top. There it may well look ridiculous and even disgusting. But see the creature at play in its native woods, its free nature finding scope and opportunity, and living in harmony with the rudeness of the scene, and instead of being absurd, it becomes graceful, and the tree seems incomplete when the creature quits it. Much the same is it with the despised plants denominated "weeds." True, if allowed to spread unchecked, many kinds establish upon farm-land a disastrous empire, that supersedes the prospective crops, strangling the roots, twining round the stems, or mingling their pernicious seeds with the wholesome grain; but this is a fact belonging to a different class altogether from that which includes the consideration of the absolute and intrinsic beauty and usefulness of the plants. Ragwort, that covers the neglected fields with gaudy yellow, nourishes the caterpillar of a lovely butterfly that will eat no other leaf with content; — thistles, that aggravate the farmer uncareful to nip them in the beginning, supply in their seeds food for innumerable little birds, especially those of the goldfinch kind. Both plants, moreover, in vigor of growth, elegance of organization, clear brightness of color, and long continuing flow of cheerful bloom, take place with the handsomest that the profusion of nature flings abroad. We may travel many miles, and explore whole provinces, and not find a more charming plant than the crimson musk-thistle. In its native haunts and proper abiding-places (which are by the edges of green lanes, and on green and breezy downs overlooking the sea, as on that fair green hill at Clevedon, from which we look across the water to South Wales, and far away westwards towards the Atlantic) it lifts a tall and woolly stem, crowned with some half-dozen gorgeous and half-drooping crimson heads, smelling of honey and musk, and more brilliant in effect than ten thousand of the far-fetched, dear-bought, fashionable exotics in gardens.

All right-minded people thank God every day for His greater gifts and bounties ; it is doubtful if any of us remember to thank Him as steadily for the simple and *common* things of nature, which we seem to ourselves to feel as our *right,* or at all events as so much a part of the very idea of the world as to become our lawful inheritance, and thus not needing to be considered as objects of gratefulness. A thankful

spirit recognizes the goodness of God in the weakest
as well as in the strongest of things ; and to my mind
it seems that while I am thankful to Him for the
lustre of the evening sky, for food and raiment, for
the bestowal of friends, for the sustenance of hope
and faith, for the prolongation of life, — though the
heart may have sorrowful tombstones in it, — still, I
fall short and forget if I am not thankful, too, for the
sweet shape, and hue, and odor, of that sea-side this-
tle, since it possesses not only an immaculate beauty
of its own, but, associated as it is with the sound of
the waves, and with events long since passed, becomes
a keynote forever to some of the sweetest experiences
of bygone life. All things deserve such thankfulness,
the commonest as well as the grandest, for the common
ones are the heritage of the poor, given them " without money and without price," so that it is but simplest philanthropy to be glad of the presence of what
all can enjoy without cost."

These, however, are matters rather divergent from
the idea of plant-life in early spring. The appearance of
the buds of trees is without question the most reliable,
since there is a greater steadiness and exactitude in
the succession and periodicity of their vital phenomena
than occurs in very many herbaceous plants, though
to appearance the latter may be quite as regular. It
is not a little curious that the renewal of trees by
annual shoots developed from buds is a matter of

comparatively recent observation. The ancients had
a name for the buds of trees ; but it was our illustrious countryman, John Ray, a minister of the gospel
in the time of the Commonwealth, who first demonstrated scientifically that the increase of trees takes
place by means of such annual sproutings. Some of
the German naturalists regard the trunk of a tree as a
mere mass of obsolete vegetable matter, and the
annual shoots as comparable to young plants that rise
out of the earth in spring from seeds. To this view
of the matter there are, however, grave objections.
Every kind of tree has a fixed lease of life ; it is competent to acquire a given stature and a definite profile
and physiognomy, and until this has been attained, it
can hardly be said that the tree is other than a living
unity. Be this as it may, the wonderful structure of
the buds, and their prodigious powers of life, are of
the most singular and striking interest. Every bud
consists of a growing nucleus, a little heart of pith
seeking to push forwards ; and overlying it, a number
of minute leaves, mere rudiments of proper leaves,
which protect the centre from the asperities of the
weather, but yield this way and that, correspondingly
with the enlargement of the germ in the interior. By
degrees true leaves are developed, a slender shoot is
protruded, and it may be that in this there is the commencement of a large bough. The outer, rudimentary
leaves undergo no change ; they retain their places as

long as useful, then drop off. The shape and color
of the buds are no less various than their economy is
admirable. In the beech-tree they are slender and
pointed, resembling brown thorns; in the oak, solid
and amber-olive color; in the ash, of the blackness of
soot; in the lime, yellowish-red; in the horse-chestnut,
covered with abundant sticky matter. Every tree, in
a word, may be told as readily in the earliest days
of spring by its buds alone, as in summer by its
flowers, and when in full leaf, by the peculiarities of
its foliage. This is one of the great charms of the
study of nature. We have always something to fall
back upon. Every season writes the names of its
trees and plants legibly and unmistakably, but in a
different mode. We never need be at loss, since the
disappearance of one feature is the signal for another
to come into view. The buds open at very various
times in the different kinds of trees. The first to come
in leaf is the woódbine or wild honeysuckle, which is
often in nearly full foliage many weeks before others
have begun to move; the elder is also very prompt;
and soon after them come those small, green, countless
specks in the hedges, that by and by are to make the
richness of the hawthorn, and become dappled with
its crowd of odorous blossoms. Marvellous is it to
note the power they have of resisting cold. Doubt-
less their progress is checked by the advent of a
frosty night after they have commenced; the wonder,

however, is not that they should be checked, but that they should hold such an amount of inner warmth as to stand proof against the bitterness that destroys tender things from India with a touch. In severe winters many trees get quite killed ; but thousands of others prove their invulnerableness, and seem, as it were, to rise from the dead. Many and wonderful have been the miracles wrought for special moral purposes, but no miracle has ever exceeded in sweet and impressive power, that great one we all witness every spring, when that which seemed quite dead, shows life in unabated energy, and this without the visible presence of a miracle-worker.

Curious is it to note also how many of the buds prepared by a tree never come to maturity, nor even sprout. Buds are disposed so symmetrically upon the branches, that were every one of them to be pushed forth into a twig, and again produce other twigs, there would soon be an inextricable mass, utterly preventive of ventilation and the entrance of light, and the tree would die of self-suffocation. But the economy of nature provides for the premature death and destruction of an enormously large proportion, no more growing than there is ample room for, yet as many as will render the tree perfect and picturesque. So admirable is the dispensation of natural laws, evoking order out of disorder, and making what seems to be injury and loss the very means towards securing

the highest beauty and perfection. Such is everywhere a grand characteristic of the works of God:
what in our short-sightedness appears defect and
blight, is in truth a preliminary step towards the
most exquisite and perfect design.

3

CHAPTER III.

APRIL, if the season be moderately genial, is one of the most remarkable months of the English year. Winter, though it may return now and then in bitter nights, is no longer felt injuriously during the day; the east winds may blacken the poplar-flowers, and try our tempers; but spring, in defiance of all hindrances, pursues its way steadily, resolutely, and with success. Nowhere is this more beautifully shown than in the vegetation of the seeds bequeathed to the soil in the previous autumn, and which after lying in the earth apparently dead for many months, now assert their intense vitality, and lift their green blades into the air. A seed is one of the most wonderful things in the world, containing not only the first principles of the plant, but holding the power to lie, as it were, asleep until the fitting period, for the expansion of the germ, and meanwhile, withstanding influences of destruction such as totally dissolve objects that have no life in them. When we consider the exquisite minuteness of many seeds, this property becomes still more amazing. Peas, beans, and similar seeds, though by

no means the largest, are yet of immense bulk com
pared with the seeds of the orchis; and these last,
though so fine as to be scarcely visible except in a
heap, are in their turn probably as much larger than
those of the moss. There is reason to believe that in
the atmosphere are constantly floating millions upon
millions of delicate germs ; that we take these germs
into our bodies when we breathe ; that they become
embedded in every species and description of food ;
that they are associated, in a word, with every con-
ceivable substance, and are as universal in their
penetration as the light of the sun. The inexpressible
minuteness of every particular seed alone renders this
possible, and perhaps it is by the minuteness that the
indestructibility is partly insured. Seeds, accordingly,
are not to be thought of merely in the idea of those
we sow in the garden, with a view to wholesome vegetables and fragrant flowers. These form but a very
minute portion of the entire quantity ; and though
their destiny may seem more dignified, it may be ques·
tioned whether in the economy of nature the little
seeds which we never behold, do not play a part quite
as salutary and important. For in judging of nature
and its processes, we err if we think those only to be
grand and splendid which are promotive of benefit to
ourselves. Since all things have been created for the
glory of God, an equal splendor attaches to every
phenomenon and process, however trifling in our eyes,

that conduces in any way to the stability and decoration of the general fabric. These tiny seeds that float in the air, have for their special function the starting of life in places where previously there was none. The moment that any surface previously bare, becomes moistened with rain or dew, they settle upon it as bees do upon flowers. If not burned up by the sun, in a little while there is a thin green film of vegetation, and by and by is seen a colony of mosses. Hence, upon the old cottage roof, especially if it be of thatch, that sweet and rich variety of tender leaf and blossom. Every spray is the growth of a seed wafted thither by the wind. It almost seems as if the atmosphere held plants in solution, and deposited them as a chemical fluid deposits crystals.

But let us inquire what a seed is composed of; what is the constitution, or at least, the aspect of those wonderful parts which a little rain and a little sunshine can tempt into expansion, and by and by develop into a flower or tree. In its most perfect state, a seed consists of several distinct elements. Outside of all is the protecting skin, by botanists called the " testa; " when this is removed, the interior is found to consist either of two solid white halves, usually flattened upon their inner surfaces, as in a pea or bean, or it consists of a quantity of white and farinaceous matter, well represented in the flower of a grain of wheat. Look a little further, and if the seed be one of the former kind, at one extremity, uniting the two halves, is a delicate hinge ; if, on the other hand, the seed be like that of the wheat-grain, the hinge-like body is embedded in the farinaceous matter. The actual and growing part of the seed is this delicate little point that we compare to a hinge. The proper name for it is the " embryo ; " and though the remaining portion is indispensable, from this alone are developed the stem and ultimate foliage. The farinanaceous matter is termed the " albumen," and is the food of the embryo while germinating, all being consumed during the processes of growth, so that when the plant makes its appearance above ground, there is nothing left below but an empty husk. Seeds that consist of two distinct halves have their albumen wrapped up in the substance of these two pieces ; and then it usually happens that at the time of germination, the seed-leaves rise to the surface of the soil, and spread themselves horizontally. Their primary function, however, is precisely the same, as proved by the experiment of breaking or tearing them off, when the embryo almost immediately withers away. The embryo of the seed is to the plant what infant offspring is to the animal ; and this leads us to one of the most beautiful considerations of their history. Providence, in assigning duties, and conferring affections and tender sympathies, gives to the mother an inexpressible love for her offspring, and impels her to

3*

nourish and cherish it; and in order that this deep love
may exercise itself in the way most needed, gives at
the same moment the physical power of replenishing
the little life from the fountains of her bosom. This
rule, in some mode or other, holds throughout the
whole extent of organic nature; and strange as it
may seem in the first statement, is not absent even
from the plant; for the seed is the offspring, and,
though cast away, often to a long distance from the
parent, is still provided for, after the same manner as
the tiny suckling; — the embryo lies between the pair
of nutrient hemispheres, and draws from them the
support needed to its fragile existence, and which
alone it can make use of. Not until it is somewhat
grown, and has become hearty, can it feed independently on the earth and water which surround it; not
until those beautiful tints of tender green make their
appearance, can it live except on the supplies derived
immediately from the parent. The production of the
fruit or seed of a plant, though in strict agreement
with the repetition of an animal of any kind, under
the law which has its maximum in parent and child, is
thus not exactly equivalent to the birthday of the
offspring. The latter, in the plant, truly commences
with the process of germination, and may be delayed
almost indefinitely.

The farinaceous matter contained in the seed does
not nourish the embryo in the crude form in which we

find it on dissecting the seed prior to the commencement of germination; the latter process begins with the conversion of the farina into a sweet and sugary fluid, which last is the actual food of the little plant, and thus forms another point of resemblance between the growing seed and the young of the lactiparous animal. This is very familiarly illustrated in the preparation of malt from barley, which is begun by sprinkling the grain with water, then warming it from below, so as to excite growth, and as soon as the sprouts appear, increasing the heat so as to destroy life. The grain, which at first was comparatively tasteless, is by the commencement of growth rendered sweet, and the result is shown in the agreeable flavor of the malt. Phenomena like these are surely quite as wonderful as those to which we are apt to confine our admiration, as the movements of the heavenly bodies, the white tumble of the waterfall, and the roll of the sea. We scarcely notice them, perhaps; but it is on the due effectuation of the great laws and principles which are expressed in such phenomena, that the permanency and the grandeur of the world depend no less importantly. Nothing in nature is large or little, or before or after another in worth or necessity. Happy the mind that tutors itself into the recognition of the Divine wisdom, not less in the arrangements made for the growth of the minutest seed, than in the majestic operations which give us

light and darkness and the seasons! One of the
greatest privileges we enjoy in these modern times is
the perception, in some small degree, of these wonderful laws and processes. They were quite unknown
to the ancients. To them was given only the external
grandeur of the universe, — more than ample, without
doubt, to fill the soul of man with rapture for ever
and ever; and though we often speak of the " good
old times," and are half inclined to wish that our lot
had been cast with that of the patriarchs, *these* are
much more really the good times, when more is spread
out by a thousandfold for the delight of our intelli-
gence and the inspiration of our fancy; — these,
moreover, are really the *old* times, for in those that are
wrongfully so called, the world, and man, and knowledge, were not old, but very young.

It is a striking and curious fact that very few seeds
are deleterious, and that those produced by plants
decidedly poisonous are, nevertheless, in many cases,
wholesome. This is observable in the seeds of certain
plants of the gourd kind, the juices of which render
them quite unfit for human food, yet the seeds are
farinaceous and nutritious. As odor is the prime duty
of flowers, so does service for food seem the essential
attribute of fruits and seeds, and taken one with
another, in truth there are very few that can be called
traitorous. It is further remarkable that plants which
secrete poisonous matters do, in some instances, store

up the venom specially in their seeds. Of this we
have a conspicuous instance in the stone-fruit trees,
such as the peach, tho nectarine, the cherry, and the
plum. The kernels of these are in every instance
reservoirs of the deadly poison called prussic acid,
whence the pungent and very peculiar flavor. Not
that the poison is present in such plenty as to be inju-
rious to the eater of a few seeds ; but there it is, stored
up by the plum for some mysterious purpose which
man does not yet understand. What a marvellous
number of such secrets are there ! Books upon scien-
tific subjects teem with knowledge of every conceiv-
able variety, and amazingly minute and accurate, and
the author often seems to have exhausted the subject ;
yet directly we come into the presence of Nature her-
self, we find ourselves lost in perplexities, and with
ten thousand more enigmas than atoms of knowledge ;
for, compared with the undiscovered, what we *do* know
is only like a few leaves from a great forest. This is
one of the great rewards of the student of nature.
He discovers very soon that the most learned cannot
explain some of the simplest things that surround
him ; thus that there are innumerable fields which he
can traverse, if he will, as an original explorer, though
he may never be able to map them out. It is not
necessary that we should acquire this power in order
to enjoy as we go along. There is more pleasure in
the pursuit than in the acquisition ; and this, we may

be sure, is why Infinite Goodness has kept out of
man's sight, so long as he is an inhabitant of this
present world, all those grand and lovely mysteries
and ultimate facts of which our actual knowledge is
only the apparel.

April is the period when the vital energy of seeds
is, in temperate countries, most vigorously called
forth. Then the gardener deposits in the soil those
copious handfuls which in a few weeks will show
themselves in wealth of young green vegetables, and
incipient flowers. Then, as if at the sound of a
trumpet, innumerable germs of the wildings of the
field and hedgerow awake to life, and beautiful is the
spectacle after a few days of sunny warmth, when the
first heralds of the season come crowding out of the
dark ground. Many of our little spring flowers run
through the whole history of life before spring has
even commenced with many of the larger and tardy
kinds. Pretty little white-flowered cresses, that do
not care to grow taller than the breadth of one's hand,
come out in the broad acres of the cornfields in absolute myriads; others peep out of the chinks and crev-
ices of old walls, opening their square and pearly
blossoms, and ripening their miniature seed-pods,
while the stately plants in the garden are scarcely
aroused. Every season is, in fact, an epitome of all
seasons; and in a single afternoon's walk, when nature
is active, the history of the whole year is found enacted

by one plant or another. The gush of new life is most marked, nevertheless, at the period we are considering, just as autumn is emphatically marked by ripe results.

Those seeds in which two distinct halves form the great mass of the contents, are proved by this structure to belong to one of the great primary classes into which all flowering plants whatever are divisible. This is a very interesting fact to take note of when at work in the garden. While trimming our borders and plucking up the weeds, it is impossible not to be struck with the appearance of these little coupled leaves, and in observing them we unconsciously become familiar with one of the leading particulars of vegetable structure. When the seed produces a pair of seed-leaves, with an embryo, we know from that little circumstance, trifling as it may appear in itself, that should it grow to be a tree, it will have branches, and boughs, and twigs, — not like a palm-tree, which is destitute of these parts, but like an oak, or an elm, or a birch. Further, we know that the leaves and the flowers will both have a specific structure ; in a word, that the whole idea of the tree will be marked by a speciality of organic form. When, on the other hand, the seed is formed like a grain of wheat, containing a large quantity of free farina, and only *one* seed-leaf, we know that the stem will be branchless, the leaves and flowers, again, with a speciality quite different from that of the others, and that the second great primary idea of botanic form is there beginning. There is only one other kind of seed, that adverted to above, as impalpably minute, and floating about in the air in millions. This is the form produced by plants which, like ferns, are destitute of genuine flowers. Internally they are different, — they sprout in a different manner; they indicate the third and last of the great types of the world of plants. Now look to the history of the creation of trees and plants. *Three* distinct classes are enumerated by the inspired writer, and learned and pious men have been led from this circumstance to believe that at the very gateway of Holy Writ there is set forth the great principle of triplicity which science in these latter ages has demonstrated. So grandly do all things lock together! Almost the last objects we should look to for a commentary on a statement in Genesis are the sprouting seeds in April; yet in these seeds are announced differences in the plants that rise from them, that every day makes more and more obvious, and which at last seem to bear out the language that cannot err.

CHAPTER IV.

LEAVES.

THE development of the leaves of plants, is the happiest sight of spring. Flowers are rarely plentiful enough to give expression to more than a very limited space at once, and although many living creatures, birds especially, make their renewed appearance at this season, it is never with such power and with such continuousness of effect. That which is true of the little, is always, in that circumstance, representatively true of the large, and thus, what becomes so obvious after a moment's thought in respect to the spring verdure of our own country, is true in an extended sense of the whole world, at least of every part of it which produces conspicuous vegetation; — *it is the green part of plants that gives expression to the landscape.* Of course there are the grand physiognomical features, the mountains, rivers, waterfalls, and so forth; but the living beauty and appeal of the landscape, come of the particular kind of verdure that may pertain to it; in England meadows and pastures, green all the year round, — in northern Europe innumerable pine and fir trees, — along the shores of the Mediterranean, vast

· 4

numbers of the plants belonging to the great race
which includes rosemary and lavender, — in Central
America strange and uncouth forms of cactus, — at
the Cape of Good Hope, heaths and evergreens almost
countless. In no degree surprising, then, is it to find
a special season so beautifully characterized ; the idea
that gives lineaments to the whole world is that which
operates to make our April and May so sweet and refreshing.

There is considerable difference in the period at which
the leaves of trees and plants unfold. So considerable is it that almanacs have frequently been constructed in which the succession of days has been denoted
by the citation of the trees which on the days indicat-
ed, or thereabouts, expand their buds, and unfold their
leaves. Such an almanac can never be made a universal one, because differences of latitude, and diver-
sities of climate even in the same latitude, materially
affect the time of commencement. But the *sequence* is
always the same, or pretty nearly so, reminding us of
what is observable in the sky. The stars, by other
persons, are from no point of view, that is at all distant from our own, seen in exactly the same places
that *we* see them ; their places *relatively to one another*
are nevertheless, always identical. In one part of
England a tree may open its leaves by the 1st of
March, and another, alongside of it, not till April 1.
In a different part the first-named tree may be a month

later, and then the second will be a month later, like-
wise. One of the very earliest to expand its foliage,
is the balsam-poplar. The buds of this are covered
with aromatic resin, so that with the evolution of the
yellow tips, the air becomes loaded with fragrance,
and very delightful is it to perceive the presence of
the tree when going along quietly after dark, by the
richness of the odor that presently meets our nostrils.
The foliage when it first comes out, is remarkable for
its yellowness. This tint is so far from uncommon,
that it may be regarded as the normal and characteristic color of very young foliage, and of such as has
not been exposed to the full influence of the solar
light; in the balsam-poplar, however, it is specially
noticeable. After a few days, when the warmth and
brightness of the sun's rays have made themselves
felt, the leaves acquire the ordinary green hue. The
influence of light in thus giving color to vegetable
matter, is one of the most striking and beautiful operations that we can witness in surveying nature. Let
any fruit be half concealed by thick foliage, and it re-
mains pale. If a stone lie upon the lawn or anywhere
conceal the grass, on removing it, the space that was
covered is found to be destitute of true green. In a
few days, however, all is changed; the sun, like a
great magician, touches what was so pallid with his
beams of enchantment, and rich and glorious hues are
almost immediately called forth. Everywhere, in the

natural, no less than in the moral world, Light is the
great life-bringer. Without it, there is no permanent
and deep-lying beauty. Well may all nations, in all
ages, have called wisdom by the name of Light, error
and ignorance by the name of Darkness, and trans-
ferred the names of Light and Brightness, to whatever
is happy and holy.

After the balsam-poplar, and almost as soon, sometimes, perhaps, contemporaneously, come out the sycamore and the horse-chestnut. In all these early trees
there is, however, a very noticeable difference in
promptitude ; that is to say, some individuals are
many days earlier than others, so that in the same
hedgerow or plantation, while many are only preparing, here and there, one will be seen in long advance.
The buds of the sycamore are shaped like almonds,
and externally pink ; the leaves which they enclose,
are folded up like a lady's fan, and gradually flatten
out, though it is several weeks before they become
fully developed. Contrariwise, those of the horse-
chestnut are the color of mahogany, and instead of being smooth and downy, are coated with viscid matter,
the purpose of which appears to be further protection
of the contents from the cold of winter. Here, again,
the young leaves are folded up like a lady's fan ; every
fold straight and symmetrical, while in the centre is
the rudiment of that noble cluster of flowers, which
by-and-by, is to help to light up this magnificent tree

as if with an ancient alabaster lamp. Often, when I
have been peering into these buds, with their simple
and beautiful prophecy, have I thought what a pretty
likeness they present to the opening heart and soul of
a child! First come forth the little green and innocent thoughts and expressions, that excite a smile and
invite a kiss. Presently, when we scarcely expect, we
hear some old-fashioned and quaint remark, that shows
what a marvellous power is at work in that little brain,
and that beautifully prefigures and pre-signifies the
glory of the intellect that will in due time be displayed
to view. Look next at that handsome white-beam,
covered with buds that, except near towns, are whiter
than those of any other native plant. We do not
call it white-beam *tree,* simply "white-beam," since
"beam" is an old word signifying tree, as illustrated
also in the name of the horn-beam. The whiteness is
given by abundant fine cottony down upon the under
surface, — the latter being the only one exposed to
view. Here, again, the primitive condition of the
leaf, is that of the lady's fan. By this disposition of
the parts, the strong ribs are all thrown to the outside,
and the delicate tissues are protected within. There
is work for an entire spring, with those who are curious in nature's mysteries, with the various methods
in which leaves are folded before they expand. They
are always plainly seen by cutting the bud crossways
with a sharp-edged knife. Sometimes, instead of being

4*

doubled up fan-wise, the leaf is rolled up like a scroll
of paper,—a plan varied by beginning from the edges,
or beginning from the centre. Sometimes the leaf is
rolled up from the apex downwards and inwards ; and
sometimes it is doubled up in a curious way that can
be compared only to a' succession of saddles, placed
one upon the other, and with an opposite set similarly
packed together. And these peculiarities are peculiar
to their own races, so that a single one will declare to
the experienced eye, almost as much as the fully
formed leaf. Cherry and plum trees may thus be distinguished from one another, before there is a speck
of either green or white on their dark-hued boughs.

The trees that come out next are the beech, the alder, and the lime. The buds of the beech resemble
long brown thorns. If we open one of them carefully,
the rudimentary leaves may be distinctly seen,—every
leaf folded fan-wise, and completely covered with
straight white hairs, that seem atoms of the finest silk.
The brown sheaths that cover them up are also ex-
ceedingly delicate ; the sheaths that lie next the leaves
are pink, and when the foliage is pretty well opened,
present a charming contrast of colors. The silky
hairs also remain for a long time, so that new beech
leaves may be identified by their presence, especially
as a large portion form a kind of fringe to the leaf,
after the manner of the eyelashes along the eyelid.
The fair green emerald light of a young wood is with-

out parallel. No place is more lovely in spring, — that is, as soon as the leaves are tolerably out; the ground is always dry, and the grasses are usually of slender kinds, quite different from those of the meadows. It is the beech that is so celebrated by the poets, as the tree suited for carving letters and names upon. The smooth bark adapts it for this purpose, better than that of any other tree ; and not only for human and veritable writing or carving, but for a very beautiful imitative writing, produced by a minute plant of the lichen kind, called *Opegrapha.* This little plant presents itself in the shape of dark and irregular lines, so exactly simulating Hebrew or Arabic characters, that we might almost fancy them to have been inscribed by some mortal penman. Where, among the works and inventions of man, — his ingenious devices and clever adaptations, — shall we find something in which nature has not anticipated him ? Nature, fresh from the hand of God, is the storehouse at once of all grand and beautiful ideas ; and the Fine-Arts Exhibition, *beforehand,* of everything that human skill contrives.

After these come the oak and the ash, the former with innumerable buds of an amber-brown, and by no means remarkably large; the other, with buds of a sooty-black color, and found chiefly towards the extremities of the twigs. The ash is one of the late risers ; seldom green all over until June, and hence, along with the mulberry, and some other trees that are very slow, made by the ancients the emblem of prudence. When they come in leaf, all danger of late spring frosts is considered to be gone by, and the tree is safe from damage. In looking for the buds of the ash, we cannot fail to be struck (in most of the trees) with the very odd appearance of the flowers. These, while young, resemble clusters of ripe blackberries; afterwards they open out into branching sprays of a peculiar blackish-olive color. In structure they are the simplest known to occur among trees. The stamens and the pistil (the parts which produce the seed), that in other plants are protected so carefully, here are left without any fence whatever; yet the tree never fails to be covered plentifully with ripe seed, as though independent of all asperities of weather, and with power to triumph over every hindrance and deficiency. These exceptions to the usual order of things in nature, form one of its most striking characteristics, and are more wonderful, it appears to me, as an illustration of the Divine wisdom, than even the method and symmetry with which they stand in such powerful contrast. They show, as it were, with such a grand independence, so self-containedly, that Infinite Wisdom, though it has chosen to construct the great mass of nature according to a given plan, is yet quite us much at home with plans and arrangements quite the opposite; and that what we suppose to be the necessities and positive requirements of things are the necessities only of the individuals in which we behold them. Step a little further, and some *other* thing dispenses with them, and yet flourishes, and is as grand and comely in default as if in possession; and that which we fancied to be the law, is shown to be only one of the ways in which a higher and greater law that we cannot reach to, is effectuated. The apparent inconsistencies of nature all meet under some higher synthesis of order which includes both the common and usual thing, and that which to our dim eyesight, seems the exception or the contravention.

The leaves of herbaceous plants make their appearance according to a similarly definite sequence. We do not notice them because they are so near the surface of the ground, whereas the branches of the trees are elevated high in air. Sweet is the spectacle on a warm afternoon in April, when we wander down by some trotting burn, where early primroses, daffodils, and anemones mix their fantasy; where the first violets seem blue eyes, and the lustrous coltsfolt glows in rays of yellow gold; — sweet is it to note the little leaves of a thousand summer-plants that, not behind time, for it is their nature to take their turn, but that in all promptitude are creeping out of the soil, spreading like green lace among the taller ones, and making green stars that have buds for the living centre. The colors, too, are as varied as those of the leaves of

trees, and usually more lucid. While in this young
state, the leaves of herbaceous plants often supply remarkably good illustrations of some parts of the internal structure. The tissues are open and succulent;
the general substance is more transparent than later
on, and the skin allows of our seeing those delicate
little openings, called "stomates," through which
moisture is transfused, and communication maintained
with the atmosphere. Leaves, in their composition,
are threefold. First, there is a delicate skeleton,
composed of fibres of woody matter, with sap and air-
vessels running alongside; the interstices of the skeleton or general framework, are filled up with green
pulp, formed of innumerable cells containing fluid;
and over the whole, on both sides of the leaf, is
spread a transparent skin, that serves to protect the
tender subjacent parts. In itself, the substance of
the leaf is colorless. The delicious and varied greens,
the deep-hued spots, the variegations, the bands,
lines, and so forth, of different hues that it presents,
come wholly of the sap contained in the cells, which
assumes one color or another when the sun shines
upon it, according to the matter secreted in it, by the
vital economy of the plant. The same is true of the
petals of flowers. The tissue itself, is totally devoid
of color. The blue, scarlet, and yellow, come of the
deposit in the respective cells of fluid, competent to
acquire those tints when acted upon by the solar ray.

How some cells should have power to secrete fluid that shall take a definite color, and no other, is one of those mysteries which at present seem quite beyond our ken. The day may come when it will be known. Happily nature is full of such enigmas. They allure us onwards, for to the true student of nature, a mystery is something to be unriddled, just as to the true worker a " difficulty " is something that has to be surmounted. It is well that we are surrounded by things seemingly inscrutable. Enterprise and imagination are alike invigorated by them. The amount of our consciousness of the unfathomed, is a capital test of our condition, for if we cease to feel the weight of mystery, we are ceasing to improve. To be satisfied with things as they seem to be, and to have no care or curiosity as to their nature and significance, is to be stranded like a ship upon the shore. Life is active in our own souls in precisely the degree that we hear it uttering itself in a thousand languages outside.

CHAPTER V.

THE STRUCTURE OF FLOWERS.

THOUGH a flower, like a sea-shell, seems one of the simplest things in nature, there are few objects that, narrowly looked at, prove to be organized more elaborately. To produce the flower, is the aim and effort of all the vital energies of the plant, from the moment when it first creeps out of the earth as a tiny seedling. For the sake of the flower the roots take up nourishment from the soil, and the leaves, that wonderful and invisible food with which the atmosphere is charged on their behalf. For the sake of the flower the plant struggles with the asperities of the winter, and responds, in the unspoken gratitude of a beautiful life, to the sweet influences and sympathies of the summer. All its energies and activities have relation to the flower as their final issue, and well may the petals be so lustrous, the odor so ravishing and balmy.

The root, the stem and branches, and the foliage or leaves of a plant, are concerned solely in the nutrition of the individual; they prepare and consolidate the material of which the general fabric is composed;

they maintain it in its position in the ground, and enable it to withstand the storm and tempest. The successive and regular development of these parts, season after season, constitutes the *growth* of the plant; and their unceasing self-employment in various vital processes, gives us those hugh masses of woody fibre that constitute timber; the innumerable secretions that we extract from plants, in the form of oils, sugar, starch, and so forth; those, also, that render the bark and the leaves of different species valuable, either to the physician, when he would prescribe medicine, or to the dyer, when he would give color to the products of the loom. The functions of these parts — the root, stem, and leaves — are thus purely of a local or personal kind; they have no direct relation to the perpetuating of the plant, and hence they are distinguished by botanists under the name of " nutritive." Very different are the functions and the purposes of the flower. Here, instead of the aim and expenditure being concentrated to the well-being of the individual, the design has reference to the race. The flower is produced, in other words, not so much for the good of the plant which it so much ornaments, as for the sake of the species; and thus, derivatively, for the beauty and verdure of the earth. Hence the parts of which it is composed are technically distinguished as "reproductive." All plants die some time. Every living thing has its lease of existence.

5

Some forms run through their little span in a few weeks, perhaps in a few days; others endure for months, years, ages; yea, scores of ages, as happens with those gigantic American trees that seem appointed to watch the rise and fall of nations; but all appear to have a definite period assigned to them, or at all events, a maximum of perfection, after attaining which they decay with less or greater quickness. This is the primary reason why flowers exist. Since plants die, as other things do, sooner or later, unless there were special arrangements made for their renewal, the earth, in a few centuries at furthest, would become bare of vegetation, and the surface be like those dreary plains of sand in northern Africa, which a fanciful author thinks are the exhausted seats of the world's first life. It is not that the plant has a lease like that which the landlord determines with his tenant. Such is not the case either, with the lease of life in animals. The idea of the lease in both classes of nature, is that of a certain relative length of life, which may be abridged or extended according to circumstances, but the term of which is as certainly fixed, as that of the summer, which may be cut short by cold rains and early frosts, or last with a calm, sweet glow of warmth and lovely hues, till it is time almost, to think of Christmas. Coming to this inevitable termination, seeds must be produced in order that a new generation shall arise; the seed is the product of the flower, and the flower is thus the fore-runner, and at the same moment, the repairer, of decay.

This matter of the lease of life, is full of pleasing suggestions, and involves the consideration of innumerable facts. It is worth noting that in plants, as in man, life consists of three great periods, two of which, the first and last, God keeps in His own hands, disposing them after His own wisdom, while the third or intermediate one, is left for man to deal with, or at least coöperatively. The first period is infancy and youth, which cannot be abided in by any man, or how many would stay there forever! The last is old age, when, having reached the crest of the mountain, and the valley of the dark shadow lies dimly below, with all our effort we cannot help sliding thither. The middle one is that glorious period, when, full in stature, and enriched with all good gifts, we feel and relish the splendors of life ; — this one it is allotted us to lengthen out almost as we please, carrying freshness of thought and feeling, which are *youth*, past as as many birthdays as suffice to see a chestnut grow from a sapling into a forest patrician. This middle period every man holds comparatively in his own power. Giving his soul to wisdom and manly affections, he finds therein the *elixir vitæ* that the alchemists sought in vain; and though the third and concluding period comes to him not less certainly than to all others, it is brief and serene.

Flowers, then, are reproductive or seed-producing apparatus. They are ordinarily so lovely, so varied, so rich, so alluring, because of the dignity of their office. That in nature which has noble duties assigned it, is always embellished in a way that befits the noble purpose. Men do just the same when they place a crown of gold and a purple robe upon that one of themselves that they choose for king. To maintain the living flow of beauty of tree and flower upon the earth, is as much grander than simply to grow stout and leafy, as it is for a man to seek to delight and illuminate those around him, (so far as the munificence of God gives power,) instead of simply seeking his own profit. This is recognized accordingly in the *forms* of flowers, so exquisitely symmetrical in the delicacy of their tints, and in their lavish profusion, all of which become characteristics of honor and dignity to the plant, and mark the period of its highest value in return. Taking the flower to pieces, it is found, when in its most perfectly developed condition, to consist of four distinct sets of parts, every one of which has its own especial province. Look at a rosebud when just about to open, and it is wholly green. The green portion is the "calyx," or chalice, enclosing everything else, falling back when the contents are ready to burst forth, and remaining usually till the seed is ripe, though sometimes cast to the ground. It is a beautiful sight when the component pieces of the calyx begin to separate, opening like the sweet, soft eyelids of a little child, when it looks forth again after the peace of its scarcely-breathing slumber. With the first slight push from within, the green separates, and a line of crimson gleams in the space; by degrees the fissure widens, and in due time, like the opening of the portals of some glorious temple, all is thrown back, and the petals stream out in their matchless hue. The same with the poppy. Two concave leaves clasp the inner secrets in their green embrace, and the bud hangs with its head to the earth until all is ready. Then the line of crimson announces what is coming, and great scarlet petals, that flaunt like banners, are disclosed. Similarly the peony unpacks great round green balls, and every other plant its peculiar cluster or bunch, — everything creeping out of this "calyx," in the first place, and seldom discarding it, though the special utility has ceased. Sometimes the calyx, instead of being green, is highly colored, emulating the most dainty parts of the interior of the flower. This happens in the common fuchsia, which has a calyx of four pieces, coherent by their edges, so as to form an oval and crimson bag, which yields to the pressure of the fingers. In other plants, again, the calyx is extremely minute, so small as to be incapable of giving adequate protection ; — then the protective purpose is subserved by special leafy coverings called "bracts," as we may plainly see in

5*

the great purple and cone-like buds of the rhododen-
dron, which are built up wholly of bracts and flowers
within.

Next in order of position to the calyx, come the
"petals." These are the parts which it is usual to
call the "leaves" of the flower; but the term is in-
correct. Leaves are the green organs which prepare
the food of the plant, and in the aggregate constitute
the foliage. These colored portions are of totally different function, and in texture and substance, are also
quite unlike. They should no more be called
"leaves," than fingers should be called toes, but
always be denominated by their special name of
"petals." Collectively they constitute the "corolla," literally the little crown, *i. e.* as signifying that
the expansion of the flower is like the placing of the
diadem on the brows of a monarch. There may be a
literal truthfulness in the name as well, depending on
the resemblance of the circular cup, found in certain
descriptions of flowers, to the golden circlet that
forms the essential portion of a crown or coronet.
But names such as these have generally an inner and
higher meaning. They were imposed by men in the
beginning from a better ground than simple compari-
son; they sprang from that intuitive perception of
the original harmonies of things, in which all the best
and most living part of language, finds its begininng
and its explanation. The corolla, like the calyx, is sim-

ply protective, or, at all events, only auxiliary to the main intent. Just as the calyx wraps up all that lies interior to it, so do the petals enfold still more interior parts, those, namely, which are directly concerned in preparing the seed, and which botanists call the stamens and the pistil. The pistil is that slender column in the very heart of the flower, which has for its pedestal the rudimentary seed-pod ; the stamens are the delicate pillars which stand around it, every one of them tipped with a little bead-like head. Both parts vary considerably in number and size. Sometimes there is but a single stamen ; sometimes there are not less than five hundred stamens, as happens in the Rose-of-Sharon. The same is the case with the pistils, which vary in number from one, which is most usual, up to a hundred or more, as in the strawberry. These two sets of organs, by their coöperation, give origin to the seed. The rudiment of the latter is contained in the ovary, which forms the base of the pistil ; the stamens, for their part, discharge a fine powder, called the pollen, which being received upon the upper extremity of the pistil, thence has its virtue conveyed downwards into the ovary, and so communicated to the potential seed. Unless this process be effectuated, the rudiments of the seed undergo no change. They never swell in the slightest degree, remaining mere shells, and unless some portion of the ovules become fertilized, the whole of the ovary generally withers away, and drops to the ground. Ovules that have not been fertilized, are often found side by side with others that have grown into perfect seeds, in the core of the apple, and in the pod of the common green pea supplied by the kitchen garden. To promote this wonderful process, and to insure its grand results, as far as possible, the most beautiful contrivances are made use of. If the flower be pendulous, the pistil is longer than the stamens, so that the fertilizing pollen, in its fall by gravitation towards the earth, shall be captured, and forced to accomplish its proper design; if the flower grow erect, the stamens are the longer parts, again that by gravitation their pollen may fall upon the crown of the pistil. While the process of fertilization is going on, the flower is apt to close at night; and when rain begins to fall, so that cold and wet may do no injury, the petals frequently draw together as if it were sunset, and form a natural pent-house over the centre. Let the sun shine bright and warm, and the petals spread out widely, forming mirrors and reflectors which cast back the light and warmth upon the delicate apparatus in the centre. Every change in the condition of the air and sky, is anticipated in the plan and the economy of the several parts, so that every possible advantage may be taken of what is favorable, and all adverse conditions be promptly guarded against. Many flowers, it is true, are so constructed as to be

incapable of self-defence. These, however, are generally of a form that is already a sufficient security, or,
they are so poised upon the stalk that the wind sways
them round, and instead of closing, they are able to
turn away their faces. So beautiful and ingenious
are the expedients that allure our interest at every
step! All this, moreover, is a part of the vesture of
the " grass of the field ; " for the apparel of the little
things of nature, whether it be lily or speedwell, primrose or golden loose-strife, that we contemplate, consists not only in the substance ; it lies as much in
their methods of life, and in the innumerable designs
for their prosperity which we behold thus effectuated,
and " if God so clothe " *them* with all tenderness of
care, ah me! may the teaching not be lost!

CHAPTER VI.

FLOWERLESS PLANTS.

WHILE plants, in their higher grades of development, are ornamented with those beautiful instruments of self-perpetuation termed flowers, others, which compose the lower grades, instead of being propagated by the agency of calyx and corolla, stamens and pistil, are in a special and popular sense *flowerless*. No plant is absolutely destitute of the means of reproducing itself; nor does any plant fail to give illustration of that wonderful twofold energy of nature which culminates in man and woman. It is true, nevertheless, that many vast tribes and races of plants, including many forms of considerable bulk and altitude, never present anything to our eyes (so long, at least, as unassisted by a microscope) that can legitmately be called a flower; while others, though they anticipate the idea of the flower, and in the most exquisite manner, do so rather in symbol than in similarity of parts and organs. Such are the lovely plants everywhere so much admired and assiduously cultivated under the name of Ferns; such, too, are sea-weeds; such, again, are mosses, and many other little plants, the pigmies of their world, passed over by incurious eyes, and uncared for by any save the botanist, but capable of supplying inexhaustible delight, and this at every season of the year. When the survey of large and showy plants has been in a measure completed, a man may go to these little flowerless plants, as into a totally new realm, begin life over again, — find that the tender ministrations of the common things of nature, even in these their most attenuated forms, are, after the love of wisdom and goodness, the true *elixir vitæ*; and discover that through their aid the surprises and wonder of the child may be renewed to him over and over again, and the more delightfully because the experience of possibly half a lifetime has supplied knowledge that renders the new facts no longer mysteries, but insights. What more exquisite, in early spring, than the spectacle of the young uncurling ferns, rising out of the earth in little coils of spongy verdure, densely clothed with brown scales, day by day taller, day by day unrolling more and more, till by and by they present the figure of a bishop's crozier, or the crook of a shepherd?

By the time that the sweet dog-rose flings out its scented cups, these coils have turned into broad, flat leaves, often with innumerable feather-like segments, but for flowers we look in vain: autumn, even another season, does not reward our expectation. Instead of flowers, the ferns produce bodies analogous only to flowers, and these are originated, usually, upon the under surface of the leaf, which they bestrew in the shape of little spangles, or embellish with broad brown furry bars. Sometimes these curious bodies, instead of being scattered upon the under surface, are disposed along the edge of the leaf, when they form a miniature braid. The particular mode of their dispersion supplies the best means of distinguishing the various kinds of ferns ; for, in ferns, as everywhere else in nature, real resemblance depends not upon superficial but upon deep-seated characteristics, and we should make great mistakes if we relied upon mere outline. Outline in ferns is usually only like apparel in human beings, which, though in some cases suggestive and even conclusive, in others may lead us astray and perhaps into peril. In the shield-fern, the seed-spangles are of a deep purplish-lead color, and disposed in double lines up the centre of its countless leaflets ; in the *Oreopteris*, or mountain-fern, the spangles run like yellow beads around the edge, following every inden-tation and delicate curve, just as the little waves, at high water, find their way into every creek, and kiss the great round pebbles they are not strong enough to encircle. Contrariwise, in the hart's-tongue, instead of spangles, we have long lines of tawny felt, that strike obliquely from the mid-vein of the leaf away to the edge ; and in the maiden-hair, delicate little semicircles, that break the otherwise even margin, and remind us of the undulated edges of certain sea-shells. These masses of seed-material, whatever their shape and position, are technically termed the *sori*, and always spoken of in the plural. A separate one, did we care to remove or isolate it, would be a *sorus*. Examined with a tolerable microscope, every " sorus " is found to consist of a heap of minute boxes, perfectly globular in form, and capable of opening into two halves, after the manner of a bivalve, such as the cockle. The opening is not effected, however, by means of a hinge, but by means of an elastic spring, which is curved half way round it while the box is young, but subsequently straightens itself, and forces the box open. The boxes, technically termed the " thecæ," contain quantities of golden-colored atoms; these are shaken out when the " thecæ " burst, and of their growth come in due time the new fern-plants. Yet they are not seeds, any more than the sori are flowers ! Ferns are, in respect of their reproduction, elaborately unique. No plants exhibit so truly wonderful an economy; they make imagination true, alike in their diversity, and in the mysteries of their life ; and it seems but fitting in so strangely-beautiful a race, that they should be contemporaneous nearly with Time, so far as registered by fossils and by living nature. For in the " great stone book " of nature, as the crust of the earth has well been designated, few records of the

6

infinite past carry the mind back to periods prior to
those when ferns existed, or, at all events, plants of
the fern idea; and in the green lace of their delicate
leaves, reappearing so sweetly year by year, now,
after a thousand ages of heritage of perfect beauty,
they are youthful and fresh as ever, and seem to
announce themselves immortal. Ferns existed in the
earliest ages of the world's history, long before man
was ushered upon the scene. Their race has seen the
rise and fall of empires, the birth and decease of
countless generations. Like the stars, in whose self-
same light they grew and flourished, they seem an
integral part of the glorious system we call our own,
and in the middle of which we live. I do not know a
more grand and exalting thought in connection with
external nature than when on a fair summer's evening,
in a country lane, while it is yet too light for the stars,
but the planets peer forth like loving eyes, we look
at these green ferns, so old and yet so young, then at
those "diamonds in the sky," so young and yet so
old, new-born and yet so ancient, and compare their
antiquity, pondering that before man was, that same
soft lustre came streaming down on their ancestry of
verdure, and that when our little lives have run their
length, and we have dropped back into the dust of mother earth, still will stream hitherward that inextinguish-
able brightness, still will these tender leaves rejoice in
their innocent life. It is when in the silent contem-

plation of these grand and awful things that, perhaps
more powerfully than at any other time, we hear, as
the little lad in the temple heard the voice, while Eli
slept, — "Have I been so long with thee, Philip, and
thou hast not known me?" These things seem moreover to waken up the reverent soul more acutely than
indoor didactics, and therefore is it good to seek their
presence, not neglecting temporal and immediate
duties and responsibilities, but in the intervals of
duty going amid them and beneath them to be re-
freshed. Fossil ferns, of the kind referred to, are
supplied by every coal-pit, — not from that portion
of the coal which is best adapted for fuel, but from
the shaly portions which lie externally to it.

Returning to the seeming seeds of ferns, which, as
we have said, are yet *not* their seeds, we have next to
ask, what then are they? If we sprinkle them upon
a piece of tile, and keep the surface moistened, in due
time the seed-like atom begins to grow, and a minute
green plate is developed. Underneath and upon the
edges of this are produced organs that execute the
functions of stamens and pistil; an actual germ is
ripened almost in the substance of the little plate, and
from this arises the new fern. The sorus on the original fern-leaf is thus a branch in miniature; every
theca in its turn is a cluster or bunch of flower-buds
in miniature, the theca itself bearing some analogy to
the white sheath that encloses the flower-buds of a

narcissus; while every seed-like body is in reality a
representative flower-bud, which only expands after
being cast away from the parent, and develops the
true seed at a distance from it. This wonderful
process, it may be well to repeat, is unique among
plants, so far at least as known, and gives the race a
most striking individuality.

Next in familiarity to the ferns are the Mosses, —
those delicate little velvety or lace-like plants that
spread themselves over the bark of old trees, on moist
rocks, upon hedge-banks, on old cottage-roofs, espe-
pecially if composed of thatch, — that grow, in fact,
in almost every habitat that can support life. The
flowers of these, though extremely minute, can be
made out much more readily than those of ferns.
While the plant still seems no more than a tuft of
minute leaves, deep down amid the recesses of the
foliage there are developed tiny organs analogous to
stamens and pistil; the latter, on being fertilized, is
elevated upon a stalk as fine as hair, and then we get
those pretty little capsules, fat and green, or ruddy
and half pendulous, that show so conspicuously to the
observant eye in early spring. They are more like
choice flowers than like seed-pods, and are so organized as to exhibit some of the phenomena that pertain
to the very highest races of plants. The rim of the
capsule is set round, for instance, with delicate and
movable rays, resembling the white border of a daisy,

and exhibiting the same kind of sensitiveness to changes of the atmosphere, — that is to say, closing when it is moist and clouded, opening out flat when the sun shines and the heaven's breath smells wooingly. The seeds are impalpably minute, and float in the air, like the winged ones of thistles, but of course invisibly. When carried or blown against any moist and shady surface, they cling to it, and commence active life, and in a little while make it seem as if a coat of bright green paint had been overspread there. Mosses, independently of the beauty of their little capsules, are plants that abound in curious interest. There is good reason to believe that the plant referred to under the name of the "hyssop that springeth out of the wall," was one of their number. Since hyssop, properly so called, is neither an inhabitant of walls, nor so remarkable for diminutiveness as to form nearly so suitable a contrast to the cedar of Lebanon, it is pretty certain, at all events, that the allusion is made to something else ; and the positive arguments are all in favor of its having been mosses that Solomon spoke of. Mosses, like ivy - and wall-flowers, consecrate themselves to the ruin. The time-worn castle, the roofless abbey, are favorite abodes with them. They love also to grow in rural graveyards, and may often be seen filling up the deep-cut letters of the epitaph. They are too small for the posy or bouquet, — not impressive enough for the flower-garden ; they seem

6*

marked out by nature for elegant and sacred purposes
that shall be all their own, and these they never fail
to fulfil. Nature never forgets either her festivities
or her tender sympathies. When spring and summer
come, the chaplets are always ready, beautiful as
gladness, and dipped in odors ; when there is anything
sad or solemn, she is ready again, and with a smile
that gives a poetical side even to death.

Sea-weeds, those lovely and fragile forms tossed
upon the sands from the country of the mermaids, —
or, in their larger and stouter forms, hanging in black
tapestry from the water-beaten cliff, — sea-weeds,
again, are flowerless plants. Yet they reproduce
themselves as regularly as those that bear stamens
and pistil. Sea-weeds differ from all other plants in
the complete fusion of all their parts into an homo-
geneous mass. There is no distinction of root, stem,
and leaves ; any and every portion is a miniature of
the whole. True, they hold fast to rocks, and beautiful is the spectacle when the tide is at its full, and the
white-bubbled waves come lashing and surging, and
the long black thongs and branches, with their great
vesicles, float and toss in the water like a strong
swimmer in his merriment, secure all the while in
their living anchorage : but the attachment is still not
that of a root, but simply of a powerful adhesion. If
anything can be compared to foliage, it will be the
slender and thread-like portions ; but the functions

even of these are not such as pertain to proper leaves, the plant taking up nutriment in every part alike. The parts analogous to flowers are contained in smaller vesicles, found chiefly near the extremities of the branches. They are too delicate to be distinguished without the aid of the microscope, but then excite the liveliest astonishment, so lively and so novel are the phenomena by which their energy is manifested. Fresh water contains plants of much the same general character as the weeds of the sea. These are usually *green*, never purple or rose-colored, like many of those that live in salt water ; and their structure is in many cases much more simple. Every pond, and stream, and fountain pool contains abundance of light green cloudy matter, which when carefully taken out and diffused in a basin of water, is found to consist of thin threads, finer than the finest silk. Examined with the microscope these threads present the appearance of necklaces, consisting of numerous oblong cells or beads joined end to end. A very common species, called by botanists *Zygnéma*, illustrates the process of seed-production in the most striking manner. The cells are filled with dark-colored granules, and as soon as these are ready, the cells of adjacent threads unite by their surfaces ; a passage is opened from one into the other ; the contents are all transfused ; and the cells that are made the receptacle of this commixture become as it were the seed-pods. So marvellous are the disclosures of the microscope !

It was said by La Place that certain discoveries in mathematics had lengthened the life of the astronomer, by enabling him to realize new privileges and new delights. As truly may it be said that the invention of the microscope has lengthened the life of the physiologist. It is more than foreign travel, — it takes us not only into countries that we have read of, and seen pictures of, but into realms peopled with undiscovered wonder, and has the additional enchantment of bearing us ever onwards, by making us feel that though we can never get into the centre of things, we may yet be always approaching nearer, and witnessing new miracles. There is as much, perhaps vastly more, in the infinite little for man to learn and to be charmed with, as he may find, or than he can grasp, perhaps, in the infinite great and distant. Man stands midway between two worlds in more senses than one. If he can look backwards and forwards as regards time, and feel equidistant alike from beginning and end, so may he, by the help of those wonderful instruments, the telescope and the microscope, perceive himself to be midway as to his place in nature.

CHAPTER VII.

FRUITS AND AUTUMNAL DAYS.

THE fruit of a plant is the portion to the development of which all activities have been dedicated. The root, the leaf, the flower, have all wrought to the furtherance of this grand intention ; and the nearer the time has come for the appearance of the fruit, the more beautiful and alluring has the aspect of the plant become. There is in this a wonderful and most exquisite analogy with the history of animal life. What the fruit is to the plant, offspring is to the creature ; and hence we find that it is towards pairing-time that birds become vocal, and dressed with gayer plumage, and that the sweet ingenuities of their little architecture begin to show themselves in the hedges, beneath the eaves of our houses, and in the innumerable quiet places where nests and fledglings are supposed to be secure, — too often, alas ! the contrary. Hence, too, we find that, at the corresponding period in their lives, fishes become more brilliant, their tinted scales gleaming with unaccustomed colors as the swift fins push the water aside, and the sunshine falls on them slantingly. Hence, too, we find that the insect, when it is about to enter on the final stage of its chequered life, and bequeathe a crowd of tiny eggs to the prospective seasons, — we find, I say, that it then assumes those glorious wings that have made the butterfly the type with many a theologian of the new state after the resurrection. The egg-state of the insect is almost negative; in the grub or caterpillar phase of its life few look on it except with disrelish; in the chrysalis phase it is again passive, and seems to have dropped out of the ranks of living things. This phase emerged from, and the wings assumed, or rather *disclosed*, — for they were already in the chrysalis, though imperfect, — the little creature mounts into the air, chases every fancy that impels to another flower, sips a little honey, sports with its gay companions, and displays those exquisitely beautiful attitudes that art feels glad to immortalize. All this is premonitory of the pairing-time of the insect, which is thus in close conformity, as to incidents and circumstances, with that of the bird; while both series of events and beautiful spectacles correspond again with those that are manifested in the plant when the period approaches for the ripening of the seed, which is only another name for the elaboration of the fruit. By the time the plant begins to blossom, it has ordinarily acquired its perfect physiognomy, — not necessarily the full stature it is capable of reaching, but it shows the perfection of its general profile. The idea of the plant is there, though it may be only in miniature, as happens with young forest-trees, and often with young trees that are grown for their economic value in gardens and orchards ; for every species of plant has a configuration of its own ; — is built, so to speak, upon a definite and prescribed model, the dimensions of which may be enlarged, and prodigiously so, as years roll over the world, but which is never materially departed from. The poplar, that when full grown towers above most of the surrounding trees, shooting up vertically, yet withal so unsociably, and giving the same pleasant idea in the landscape that spires and columns do in the view of a large town from the hillside, — this tree, in its youngest state, is a kind of living photograph of its tallest ancestor, presenting all the characters that in the mature one are merely repeated, without being in any degree diversified. This general figure and physiognomy are realized by the time the plant begins to show its blossoms. The latter seldom appear before, unless under some constraining influence that hurries the plant to early death, as when it grows on some dry wayside, or on some scorched and sunburnt cliff, whence every particle of moisture is rapidly evaporating, and then, it is true, an effort is often made to produce a seed rather than die childless. For here, as in other ways, the plant gives us a profound and beautiful lesson. It is not *dying* that is dreadful, or to be looked on with

dismay, but dying without having *lived*, *i. e.*
without having lived to some good purpose, however
slight, so that the best was done that could be. Plants
under cultivation are often reluctant to produce blossoms. Year after year they unfold abundance of green
leaves, and as " foliage-plants," command our admiration ; but we are never gratified with the sight of a
flower. The plan generally adopted with such plants
is to starve them in some way ; checking the exuberance of growth, alarming them, as it were, with the
fear of being destroyed, when they forthwith make
efforts to produce flowers, so that they may leave at
least a representative of their race. From Mexico
was brought, a few years since, a magnificent plant,
named, in compliment to a celebrated French Admiral,
Bugainvillea. This is so unwilling to blossom, that in
our hothouses it often presents no more than a tapestry of dark-green leaves. The expedient that is said
to have answered best in the way of persuasion, is to
lay bare the roots, and bring one of the hot water
pipes in contact with them, whereby they are in a
degree dried up. Then the grand lilac clusters make
their appearance, every flower seeming a design in
muslin rather than the work of a plant.

While the physiognomy and general idea of the
plant are thus attained as an antecedent of the fruit,
the leaves also acquire their perfection. The youngest leaves are often quite unlike those that come after-

wards, just as certain features in the human face are
in infancy merely forehead and nose, and we have to
wait till the 'teens for their established shape. This
is very prettily marked in seedling trees, such as
the sycamore. The full-grown leaf is shaped like that
of the vine. Three great broad angles divide the
surface, and these are again distributed into smaller
angles. But while young, the leaf is simply oval and
attenuated, the angles not appearing till the tree is
many months old, — if tree it can be called before a
winter has passed over it. In other respects the
foliage of plants is often inferior while young, compared with its condition when the flowers are about to
burst. When the flowers are first opening, the peculiar properties and qualities of the leaves are also most
remarkable. If it be bitterness, or aroma, or adapt-
edness for use in medicine that characterizes the
leaves, this is the period when they should be col-
lected. Before the flowers appear, they are comparatively deficient; in the old age of the plant they have
nothing to yield. Hence it may be observed that in
the herb markets, the great sheaves thither brought
of sanctuary, betony, blushwort, and a score of
others, are always loaded at one extremity with their
withered corollas. Hence, too, in districts where
there is great faith in herb-teas of different kinds,
plants valued for such service, and that reproduce
themselves annually from seed, in time become very

7

scarce and almost extinct. They are torn out of the ground at the most critical time, and the seed that should renew them when spring returns with its encouraging showers, is forbidden to ripen. The botanists who carry off handfuls of specimens as so many trophies of their explorations, have far less to answer for than the herb-gatherers, who it would scarcely be too severe to call the locusts of modern ages.

Last in the order of preparation for the fruit comes the glow and the grace of the flower. When this makes its appearance, it is the aurora of the plant's fecundity, — much, it is true, may be repressed by blight and chill, just as a heavenly morning-dawn in early summer, that cheers the heart of the little lark, is not seldom changed into cold and gray, by winds that bring unwelcome clouds, — but the *intent* of the flower is that fruit shall be its sequence. Therefore the queenly and incomparable hues; therefore the odors that seem breath inherited from Eden ; therefore the forms and outlines before which the mathematician is still a child. We might be sure that some great event was near at hand, did experience not assure us that fruit would follow all this outlay, since grandeur of announcement in nature is always prophetic of something opulent to follow. In nature, the herald's trumpet is never blown in idleness or sport. It is always an illustration, once over again, of the incomparable veracity that belongs to every department of the works of God, so long as unmarred by man. The silent predictions of nature would be theme enough for a great volume. The maturity of a man's lifetime would be well devoted to the record of them and to their fulfilments, each one in turn,— text for a preacher, and theme for a Christian philosopher. Many flowers, it is true, fail in their promise ; but this is no fault of the mechanism, — it is no fault in any shape. Were every flower that opens to ripen its fruit, the strain would be greater than the plant could bear ; and, as illustrated every summer in the flower-garden, exuberance would be followed by slackness, perhaps by total sterility. Therefore the exquisite taste and wisdom, which in order not to use the Divine name too frequently, men have agreed to call " Nature," co-operate in such a way that while the decoration shall be all that the soul of man can desire in the shape of beauty, a due proportion only shall be actually permitted to serve the still higher purpose. We can suppose what would be the condition of trees and plants, were every individual to become fruitful, by imagining what society would be, were every little boy to become a Plato, and every little girl a Corinne.

The flowers gone by, the seed-pod makes its appearance. This is no *new* thing. The seed was always in the heart of the flower, as love in the heart of a woman ; it wants only sunshine to bring it into view.

And like the heart of a woman, it takes its coloring again from that which environs it. Let it be shade, and there is nothing but coldness and insipidity; let it be sunshine, and life leaps up as at the touch of an enchanter. As said in a former paper, we never see the actual beginnings of anything The fruit that shines amid the foliage lay in the flower, and was a rudiment when the petals were scarcely broken. Its appearance, like that of Spring, is only the last of a long series of preparations. All plants produce fruit of some kind or other, — not necessarily eatable fruit, any more than every person talks and acts so as to please and instruct. But in some shape or other, last year, or this year, or next year, it is to be found; and though it may not answer our ideas of fruit, as founded on grapes and apples, it is still fruit in the strict and proper sense. Acorns are the fruit of the oak quite as truly as filberts are the fruit of that well-known tree which, in its wild state, we call the hazel. " Acorn" is literally "oak-fruit," "corn" being no more than a general term for fruit, — wherefore we speak of fields of corn, as distinguished from those which supply our tables with roots and leaves. The fruit, in a word, is the ripened pistil of the flower, comprehending both shell and contents. Here, in England, we have an immense preponderance of dry and inedible fruits. In the tropics, on the contrary, those adapted for human food are very numerous, and comprise forms and flowers to which cold countries give no parallel. Such are the fruits of the thousand kinds of palm-tree, the custard-apple, the bread-fruit, and the guava. Now and then these illustrious exotics condescend to ripen their produce in our hothouses, but it seldom acquires the flavor that pertains to it in its native country. The skill of the gardener can supply warmth, but it cannot bestow *light.* "Let there be light," is the beginning everywhere, both in the moral world and in the natural. Wanting the clear, shining, and intense brightness of the tropics, the mere enjoyment of artificial heat will not suffice. Nor, for the matter of that, would light suffice to ripen them without a due proportion of warmth. The old story meets us everywhere. The head and the heart, the intellect and the affections, knowledge and good dispositions, — what are they, deprived of their correlatives? Inside the fruit is the seed. This is the last, grand, and crowning effort of the plant's existence; for in the seed lies wrapped the future one. Perhaps a mere speck, yet capable of unfolding by degrees, and absorbing from the earth and atmosphere that marvellous sustenance which, invisible to our eyes, shall yet be wrought into wood and sap, and built into great boughs and branches, till a living pillar is erected that shall withstand the shocks of ages.

7*

And thus we are brought round to the point from which we started. Every ending is a beginning; everything completed is a pedestal for something then to be commenced. All things are forever striving to begin over again. We rise every morning with new hopes, new desires, new resolves. Would that our new beginnings were always as fully replenshed as that of the plant in its little seed!

THE OAK.

TREES constitute an order of nobility; for nature has its aristocracy as well as mankind. If there be "ancient and noble" families in a nation or a community, — still older, and inheriting yet more dignity, are the families of living things by which man is encircled. He can claim no honor on the score of descent or genealogy that is not already merited by some patrician of the world of plants; and this not so much because trees are the same to-day that they were in the beginning, as by reason of their absolute excellence, their serene and invulnerable perfection.

Trees are sanitary agents in the economy of the world we live in. By the process of "assimilation," which means the abstraction of carbon from the atmosphere, in order that in due time, and through certain vital processes, it may be converted into wood and other vegetable substances, — by the process of "assimilation," we say, trees, through the medium of their leaves, preserve the air in a condition fit for human breathing. Herbaceous vegetation greatly contributes to this great end; but the result is mainly referable to arborescent plants, their size and extent of leaf-surface being so prodigiously great, when compared with that of the former kind. We little think when we inhale the fresh air, and quaff it upon the hills, like so much invisible and aerial wine, that its purity and healthfulness come of the glorious trees. But so it is. Nor have we merely the trees of our own country to think of and be thankful to. The air that wo breathe in England to-day has been purified for us perhaps a thousand miles away. If the wind blow from the north, we may be grateful to the Scandinavian birches; if from the west, it is quite possible that the magnolias of North America may have helped to strain it; if from the south, were it gifted with language, we might hear a tale of Indian palms. Every tree in nature makes itself felt in the good it does the air, — a beautiful return for the new loveliness it receives when its branches and foliage are stirred and fluttered by the breeze.

Trees supply man with every species of useful article, whether of nourishment, or of clothing, or of medicine, and with timber to construct dwellings, and to build ships with, so that even the sea shall be a highway. Not that any single kind is of utility so multiform. Fruits are supplied by some, as the olive and the fig, the cocoa-nut and the date; the delicate inner bark of the paper-mulberry furnishes the inhabitants of the South Sea Islands with materials for their simple apparel; medicines are afforded by innumerable species, and " wood " and " tree " are words almost synonymous. It would be foolish and presumptuous to say that man could not exist without trees, because, were there no such productions in existence, the Infinite Benevolence would supply his wants through some other medium. But constituted as man is, and established as Trees and their functions and properties are, it is plain that the present exquisite order and harmony of things in respect to man's welfare, are most intimately and inseparably identified with trees. Thus, that when we would consider man and his privileges, the amenities and the enjoyments that encircle life, the comforts and the ornaments of his home, we cannot possibly do so, if we would give all things their fair place, without keeping trees also constantly before the mind.

Trees are indispensable to the picturesque. A great mountain, or an extended plain, may have grandeur, though devoid of trees; and it is easy to conceive of richly-cultivated valleys, covered with crops of corn, or unrolling infinite reaches of green pasture, and at the same time without a tree, except a little one here and there, just sufficient to serve as a landmark. But in the absence of trees, none of these places could be *picturesque*, in the full and proper sense of the word. The trees break the outlines; they give variety of colors, movement also, and shad-

ows, and touch the imagination with agreeable sense
of fruitfulness ; or if they be timber and forest-trees,
with the idea of nobleness and grandeur. They are
to the landscape what living and moving people are to
the street, or to the interior of the hall or temple, —
an element that *may* be dispensed with, but at the ex-
pense of the finest and most impressive influences.
We may be overpowered by the stern and solemn
grandeur of a treeless waste, especially if it be composed of mountains ; and the sensation is one that
gives a variety not unacceptable to our experiences of
external nature ; but the scenes that come home most
closely to our sympathies, and that have a perennial
hold, are those that are enriched by the abundance of
their trees.

Poetry finds in trees no little of its sustenance.
From the most ancient poets downwards, all verses
that have immortality in them, abound more or less
with allusions to trees, finding in them either images
for the events, — both glad and sorrowful, — of human
life, or emblems, in their higher nature, of what pertains to the heart and mind. The "Language of
Flowers" would be incomplete did it not include the
"Language of Trees," since trees are adapted, by
their original and inalienable constitution, to serve as
metaphors for almost everything great and good, and
wise and beautiful, in human nature. Hence the
countless citations of trees in Holy Writ, wherein

the cedar and the fir, the vine and the olive, the palm and the fig, are a portion of the ordinary vocabulary, — not mentioned arbitrarily, or as a sportive act of the fancy, but on account of their being the absolute representatives and pictured forms in the temporal world of the high and sacred realities that belong to the invisible and eternal.

Because of these admirable attributes and characters of trees, we purpose in this series of papers to examine somewhat closely into their nature and life-history, marking out the features and physiognomy of such kinds as belong to our own island, and inquiring into the specialities that give them their several places in art and poetry. For a tree is not merely an oak, or an ash, or an elm. It has qualities for the imagination and the heart, moving men in its own way, and vindicating prerogatives that are peculiar to it. The mind of that man grows up very differently who in his youth is accustomed to contemplate oaks, than that of him whose boyhood is spent near pines and firs. Where evergreen trees prevail, and are a daily spectacle, a very different frame of mind is induced than exists where the branches are leafless throughout the winter. As the stars and planets, from the inaccessible altitude of their sweet lustre, make the heart great by the contemplation of them ; so, after the same manner, imposing and magnificent trees, whose branches, when we go beneath, seem the clouds of a green heaven, have a power of ennobling and elevating the soul, such as all who have lived among such are more or less clearly conscious of, and which is totally impossible to little ones.

In England, the trees are all of the class called "exogenous," that is to say, they have numerous and spreading branches; the leaves, when held between the eye and the light, are found to be marked in every portion by a delicate network of green lines, technically called the "veins;" and upon the outside of the trunk there is bark, which can be removed like the peel of an orange. When one of these exogenous or branching trees is cut down, or if a branch be lopped off, the exposed surface, on being smoothened horizontally, shows elegant concentric circles, surrounding a central point, which in young parts of the tree indicates a column of living pith. The concentric circles announce the age of the tree or branch, which is just as many years old in that part as there are rings. In its earliest stage, or while only in its first season of growth, the stem of the seedling tree consists only of pith and an enclosing skin. Woody matter is gradually prepared, and this becomes deposited in a layer between the pith and the skin, which latter now assumes the solidity of bark; and should the stem be cut through at Christmas, or at the end of its first year, the first of these annual rings will be plainly visible. Every successive year this process is repeated. With the opening of the leaves in spring — for it is the leaves that really effect the work — the preparation and deposit of a new layer of wood is commenced, so that by the close of the second season there are two layers ; by the close of the third season, three layers ; and so on as long as the vital lease of the tree endures. The bark is simultaneously renewed, enclosing a larger mass every year. This mode of growth is prettily illustrated in the spreading of the little wave-circles upon the surface of still water. Standing on the margin of some lovely lake or mere, and looking at the blue sky and the white clouds that are reflected in its clear bosom, how often the fairy spectacle is broken in an instant by the wing of some light bird that, skimming through the air, just touches the surface and sweeps onwards. But the effect of that touch is to cause circle after circle of tiny wavelet to move away from the spot where the touch was given, and as far as the eye can reach the beautiful phenomenon is continued. Just like this succession of wave-circles is that of the annual wood-circles of a tree, only that on the water we have only an evanescent effect, while in the tree there is new substance and solidity. The mode of gowth and the phenomena referred to are denoted by this word "exogenous," which is literally no more than "expansion outwards." Very different are the mode of growth and the internal condition of the trees called "endog-

8

enous." These show no distinction of bark, and wood, and pith ; they are destitute of branches (except in a few very curious and exceptional instances) ; and their leaves, which are inconceivably enormous to any one who has never seen leaves larger than those of English trees, are produced only upon the summit of the stem. They are chiefly represented in the illustrious tropical productions known as palm-trees, — those soul-moving emblems of the south and east, and in England are only seen in large and costly conservatories, where room can be afforded them to lift their great green pride on high. Even then we only see them as juveniles, no possible structure of glass being competent to shelter them when full grown, except in the case of some of the dwarfer kinds. It is among the exogenous trees, accordingly, that in England we find our delight. It is these which form the sweet and solitary arcades of the forest ; that are the homes or the resting-places of the birds ; that shelter us from the storm, and temper the heat of the sun ; whose trunks are embossed with tender creepers of green moss, or hidden by the activity of the innumerable and ubiquitous ivy ;— it is these that are so lovely in their youth, so venerable and patriarchal in their old age ; these that stand still in quiet dignity while we talk of fourscore as a wonderful lifetime, and for their own part watch the rise and fall even of nations. For the nature of an exogenous tree being to expand

and enlarge *externally*, there is of course no physical
limit to the diameter it may attain, or to the number
and massiveness of its boughs and branches, or to the
multiplication of its twigs and leaves; and should the
lease of life allowed it in the Divine economy be considerable, as happens with certain kinds of mimosa,
and with many of the pine and cedar kind, it may go
on growing and growing for ages, and after a thousand years be still in the full vigor of its existence.
Hence it is that the grand scriptural image acquires
such richness and force — " As the days of a tree are
the days of my people." Hundreds of trees are
standing at this moment in America, some in Califor-
nia, others in Brazil, that were alive when those words
were written, and with a grasp upon life and the earth
that seems to assure them a period of which they have
perhaps no more than passed the middle. England
possesses multitudes of endogenous *plants*, though no
endogenous *trees*. Lilies, grasses, rushes, are all
structurally of the same nature as the palm-trees,
and now and then they give us a pretty prototype of
the palm; but the *beau idéal* of the endogen, as said
before, belongs to the equinoctial regions. It is a
proud and inspiring thought for us nevertheless, that
art and the skill of the gardener allow us the sight of
them. By virtue of our hothouses and conservatories,
we who live in this age are introduced to the vegetation of every part of the world, without the trouble or

risk of departing either long or far from home. England, which stands midway between extreme cold and
extreme heat, with a surface that embodies in miniature every element and ingredient, except the volcano,
that gives variety and sublimity to the face of the
earth ; — England, through its art and science, is the
EXHIBITION of the whole world. We need but ask for
Saloon A, or Saloon B, and all that the heart can desire is displayed to view. Kew ; Chatsworth ; if we
cross the Tweed, Edinburgh ; and Dublin, if we make
our way to the green isle, show collections of palms,
among other things, that amply inform us as to their
wonderful nature. In these glorious places we see
the tropical regions as in a concave mirror, or in a
stereoscope, with the added charm that all around us
is alive.

Foremost among British trees, alike in grandeur,
utility, length of life, and amplitude of association,
stands the Oak, — that famous production which even
in the days of Homer was a time-honored proverb for
strength and endurance. "Thou," says one of his
heroes to a man who quailed, "art not made of the
oak of ancient story." * In England this noble tree
is found under many different forms, the contour, the
endurance of the foliage, the figure of the leaf and
acorn, varying considerably more than the unobserv-
ant of minute particulars would ever suppose. All the

* Odyssey xix. 163.

varieties are resolvable, however, into two principal
ones, and these two are so nearly connected by inter-
mediates, that it is probable the oak of old England is
after all very like a human face, — presented under innumerable profiles and complexions, but always and
everywhere the same good old-fashioned combination
of features that was possessed in the beginning. The
two principal forms are the wavy-leaved oak and the
flat-leaved, called respectively by men of science,
Quercus robur and *Quercus sessiliflora.* The former is
distinguished by its remarkably tortuous branches,
and the irregular disposition of the foliage, every leaf
lying in a different plane, and the whole presenting
an aspect of great massiveness. Leaf-stalk there is
scarcely any; the acorns, on the other hand, are
borne upon peduncles of several inches in length.
Individually, the leaves, as expressed in the name,
have a strong tendency to be wavy in their surface
and outline. The flat-leaved oak differs in its com-
pact form, and strong disposition to roundness; the
branches are more horizontal, the leaves lie in parallel
planes, and individually are flat, and with rather long
stalks. In spring, we may further observe that the
leaf-buds are larger; and in autumn that the acorns
are shorter and broader than in the other, and that
they are almost or totally destitute of peduncles; if
present, the peduncles are stout, not slim and delicate
as in the wave-leaved. These are distinctions very

8*

easily made out. To trace them is at once an agreeable and instructive occupation for half an hour, when we go into the country for a day's enjoyment. Nor does it end in the simple discrimination of two different things; for the wave-leaved oak has the reputation of being a more excellent tree than the other, while the flat-leaved is considered better adapted to excite ideas of the picturesque. A glorious spectacle is that of the oak in the month of April, when its amber-tinted buds stud the tree like so many jewels. They do not open hurriedly, like those of the sycamore or the horse-chestnut. From first to last, the life of the oak seems characterized by placidity. It lives so long that it can afford to be leisurely in all its movements, and at every season alike expresses dignity and calmness. In a little while, when the young leaves are half-expanded, come the flowers, though not such flowers as we use for bouquets. Nature has other ways of fashioning flowers than after the model of the rose or lily. To note them is one of the great rewards and charms of Botany, — which does not mean calling plants by Latin names, but exploring the wonderful nature of their various parts, and how exquisitely they are fitted for their several uses and destinies, and then comparing them with other forms of leaves and flowers, and discerning step by step that nature is all one song, but coming forth in countless tones, or rather like a grand Oratorio, where we never have two parts exactly alike, yet everywhere repetition and reverberation to the ear that knows how to listen. Flowers are not necessarily sumptuous, and fragrant, and brilliant-hued in order to *be* flowers. The idea of a flower implies simply an elegant mechanism for the production of seed, and that this be large or small is of no more importance than that the heavenly teachings should be printed in one kind of type or another. It is worthy of note also that the great timber-trees of the North are remarkable, as a rule, for the insignificance of their flowers. The short-lived vegetation of the field and garden seems decked with its sweet flower-brightness in compensation. Where our hearts are to be lifted up in admiration of strength and patriarchal majesty, the allurement of flowers can be dispensed with. Those of the oak, as said above, make their appearance contemporaneously with the young leaves, and under two different forms. First, there are innumerable yellowish tufts and fringes depending from near the extremities of the twigs; among them are the tips of the rudiments of the future acorns, scarcely larger than the head of a pin, and of a deep red color.

The oak is thus one of the trees in which the distinction of sex is strongly marked. All plants express, in some way or other, the omnipresence in organic nature of masculine and feminine. But it is not always palpable to the eye. Some philosophers con-

sider that where it is most plainly shown, we have a
nearer step towards perfection of structure; and on
this ground regard the oak and its congeners as far
more noble in the scale of vegetable life even than
apple-trees or vines. Acorns would never be developed from the rudiments in question, were the tasselled
fringes not to coöperate, and contrariwise the tasselled
fringes would yield no acorns. Summer aids the development; then comes serene October, and the pretty
embossed cups, round as a bubble upon the water,
holding them up awhile, as a young mother holds up
her child, cast them to the earth in kindly largess.
But although the acorns may sprout where they fall,
none grow to be even saplings beneath the shade of
the parent. Only those that get carried to a little
distance become oaks. And this has been observed
to be largely through the instrumentality of squirrels.
So beautifully are the necessities of the various realms
of nature harmonized one to the other. The little
quadruped fulfils an instinct proper and needful to its
own existence, and in so doing, contributes to the
perpetuation of the tree.

Representatively, — that is, as viewed by the light
of poetry, which means, in turn, by the keenest in-
sight of the mind, that penetrating below the surface,
and beholding the centres of things, brings out their
highest value, that is to say, their Significance, — representatively, the oak is strength, endurance, and dig-

nity, holding the same place among trees that the lion does among animals, and the eagle among birds. Hence we find it many times referred to in Scripture, and always in connection with what is understood to be permanent and enduring, — as when the tables of the law are described as having been set up against an oak, to signify that the law was given to last forever. It would be a very trifling piece of information for the dignity of Scripture to communicate, if it were no more than the bare physical fact that the tables were placed against an oak. Scripture always *means* something, — it does not simply speak. It is not a book of words, but of ideas, speaking for all time; which kind of language comes of the facts that it records being not simply literal but representative. It is literally true, without doubt, that the tables were placed against an oak; it is no less true that an oak was chosen because of its symbolic meaning for all ages. The poetical character of the oak is beautifully acknowledged again in the time-honored allusion to the defenders of our country as "hearts-of-oak." No one disputes the fact that our sailors are made of this capital material; yet how absurd the statement, if taken in any other light than that of poetry! This shows that although much that holds the form and outward show of poetry may be unmeaning and silly, the inmost and true spirit of poetry finds a response in universal human nature, and that its genuine language never needs interpreting.

www.ingramcontent.com/pod-product-compliance
Lightning Source LLC
Chambersburg PA
CBHW031441270326
41930CB00007B/825